FROM ROCK TO ROAD

by Shannon Zemlicka

Lerner Publications Company / Minneapolis

Text copyright © 2004 by Lerner Publications Company

All rights reserved. International copyright secured. No part of this book may be reproduced, stored in a retrieval system, or transmitted in any form or by any means—electronic, mechanical, photocopying, recording, or otherwise—without the prior written permission of Lerner Publications Company, except for the inclusion of brief quotations in an acknowledged review.

Lerner Publications Company
A division of Lerner Publishing Group
241 First Avenue North
Minneapolis, MN 55401 U.S.A.

Website address: www.lernerbooks.com

Library of Congress Cataloging-in-Publication Data

Zemlicka, Shannon.
    From rock to road / by Shannon Zemlicka.
      p. cm. — (Start to finish)
  Includes index.
    Summary: Briefly introduces the process by which rocks and asphalt are turned into a roadway using tractors, bulldozers, graders, and other equipment.
    ISBN: 0–8225–1391–5
    1. Roads—Design and construction—Juvenile literature. 2. Rocks—Juvenile literature. [1. Roads—Design and construction. 2. Construction equipment. 3. Rocks.]
  I. Title. II. Start to finish (Minneapolis, Minn.)
TE149.Z46 2004
625.7—dc22   2003011736

Manufactured in the United States of America
1 2 3 4 5 6 – DP – 09 08 07 06 05 04

The photographs in this book appear courtesy of: © John Deere & Co., cover, p. 13; Minnesota Department of Transportation/David R. Gonzalez, pp. 1 (top), 7, 9, 15, 17, 21; © GoodShoot/SuperStock, pp. 1 (bottom), 23; © Howard Ande, pp. 3, 11; © Kevin Fleming/CORBIS, p. 5; © Joel Stettenheim/CORBIS, p. 19.

Table of Contents

Machines dig up rocks 4

The rocks are crushed 6

The rocks are washed 8

The land is cleared 10

Machines shape the land . 12

Trucks move the rocks . . . 14

The rocks are spread 16

The rocks are sprayed . . . 18

A truck paints lines 20

Here come the cars! 22

Glossary 24

Index 24

Cars drive on roads.
How is a road built?

Machines dig up rocks.

Roads are built with rocks. Machines dig up rocks from the ground in a place called a **quarry.**

The rocks are crushed.

Another machine smashes the rocks. The rocks are crushed into small pieces.

The rocks are washed.

The pieces of rock are dusty.
Water is sprayed on the rocks.
The water washes away the dust.

The land is cleared.

A machine called a **bulldozer** clears away grass and trees. The land is ready for the road to be built. The bulldozer makes a dirt road.

Machines shape the land.

The dirt on top of the road is soft. Soft dirt will fall apart under cars and trucks. A machine called a **scraper** pushes the soft dirt off the road. A **grader** moves more dirt to make the road smooth and flat.

Trucks move the rocks.

Dump trucks move the crushed rocks from the quarry to the road. The trucks dump the rocks in big piles.

The rocks are spread.

A tractor rakes the rocks over the road. A machine called a roller rolls over the rocks to pack them down flat.

The rocks are sprayed.

A machine sprays the rocks with a sticky mix called **asphalt.** The asphalt hardens over the rocks. It makes the road strong and smooth.

A truck paints lines.

A truck paints lines on the road. The lines tell drivers where it is safe to drive.

Here come the cars!

Cars, trucks, and motorcycles zoom everywhere on roads. All roads started as rocks!

Glossary

asphalt (AS-fahlt): a sticky mix that makes the road strong and smooth

bulldozer (BUL-doh-zur): a machine that clears land

grader (GRAY-dur): a machine that moves dirt to make a road smooth

quarry (KWOHR-ee): a place where rocks are dug up

scraper (SKRAY-pur): a machine that moves soft dirt off the top of a road

Index

cars, 3, 22

crushing, 6

digging, 4

dirt, 10, 12

dust, 8

roller, 16

spraying, 8, 18

trucks, 14, 20, 22

WITHDRAWN FROM
HERRICK DISTRICT LIBRARY